RECREATIONAL PILOT AND PRIVATE PILOT KNOWLEDGE TEST GUIDE

U.S. DEPARTMENT OF TRANSPORTATION
FEDERAL AVIATION ADMINISTRATION
Flight Standards Service

Preface

FAA-G-8082-17, Recreational Pilot and Private Pilot Knowledge Test Guide, provides information for obtaining authorization to take the recreational pilot or private pilot knowledge tests. Appendix 1 provides lists of reference materials and subject matter knowledge codes, and a list of computer testing designees (CTD's).

Changes to the subject matter knowledge codes will be published in AC 60-25, Reference Materials and Subject Matter Knowledge Codes for Airman Knowledge Testing.

The current Flight Standards Service airman training and testing material, questions banks, and subject matter knowledge codes for all airman certificates and ratings can be obtained from the Regulatory Support Division, AFS-600, home page on the Internet.

The Regulatory Support Division's Internet address is: http://www.mmac.jccbi.gov/afs/afs600

FAA-G-8082-17 supersedes Advisory Circular (AC) 61-117, dated 1995, and can be purchased from the Superintendent of Documents, U.S. Government Printing Office (GPO), Washington, DC 20402-9325, or from U.S. Government Bookstores located in major cities throughout the United States. For an explanation of why the Recreational Pilot and Private Pilot Knowledge Test Guide was taken out of the AC system, refer to AC 60-29, Renumbering of Airman Training and Testing Publications.

Comments regarding this guide should be sent to the Federal Aviation Administration, Airman Testing Standards Branch, AFS-630, Attn: Private Pilot Certification Area Manager, P.O. Box 25082, Oklahoma City, OK 73125.

CONTENTS

APPENDIX 1

Introduction

What is required to become a skilled and effective recreational pilot or private pilot? Although some individuals possess more knowledge and skills than others, no one is a natural-born pilot. Competent pilots become so through study, training, and experience.

This knowledge test guide will answer most of your questions about taking a recreational pilot or private pilot knowledge test by covering the following areas: knowledge test eligibility requirements; knowledge areas on the tests; descriptions of the tests; process for taking a knowledge test; use of test aids and materials; cheating or other unauthorized conduct; validity of Airman Test Reports; and retesting procedures.

This guide will help in preparing you to take one or all of the following tests

- Recreational Pilot—Airplane
- Recreational Pilot—Rotorcraft/Helicopter
- Recreational Pilot—Rotorcraft/Gyroplane
- Private Pilot—Airplane/Recreational Pilot–Transition
- Private Pilot—Helicopter/Recreational Pilot–Transition
- Private Pilot—Gyroplane/Recreational Pilot–Transition
- Private Pilot—Airplane
- Private Pilot—Rotorcraft/Helicopter
- Private Pilot—Rotorcraft/Gyroplane
- Private Pilot—Glider
- Private Pilot—Balloon–Hot Air
- Private Pilot—Balloon–Gas
- Private Pilot—Lighter-Than-Air–Airship

This guide is not offered as an easy way to obtain the necessary information for passing the knowledge tests. Rather, the intent of this guide is to define and narrow the field of study to the required knowledge areas included in the tests.

Knowledge Test Eligibility Requirements

If you are pursuing a recreational pilot or private pilot certificate, you should review: Title 14 of the Code of Federal Regulations (14 CFR) part 61, section 61.23, Medical Certificates: Requirements and Duration; 14 CFR section 61.35, Knowledge Test: Prerequisites and Passing Grades; and 14 CFR section 61.83, Eligibility Requirements for Student Pilot, for detailed information pertaining to prerequisites and eligibility.

If you are pursuing a recreational pilot certificate, you should review 14 CFR section 61.96, Applicability and Eligibility Requirements: General, for additional detailed information pertaining to eligibility.

If you are pursuing a private pilot certificate, you should review 14 CFR section 61.103, Eligibility Requirements: General, for additional detailed information pertaining to eligibility.

Knowledge Areas on the Tests

Recreational pilot and private pilot tests are comprehensive because they must test your knowledge in many subject areas.

If you are pursuing a recreational pilot certificate or added rating, you should review 14 CFR section 61.97, Aeronautical Knowledge, for the knowledge areas on the tests.

If you are pursuing a private pilot certificate or added rating, you should review 14 CFR section 61.105, Aeronautical Knowledge, for the knowledge areas on the tests.

Descriptions of the Tests

All test questions are the objective, multiple-choice type. Each question can be answered by the selection of a single response. Each test question is independent of other questions; therefore, a correct response to one does not depend upon, or influence, the correct response to another.

The following tests each contain 50 questions, and you are allowed 2 hours to complete each test.

- Recreational Pilot—Airplane
- Recreational Pilot—Rotorcraft/Helicopter
- Recreational Pilot—Rotorcraft/Gyroplane

The following tests each contain 30 questions, and you are allowed 1.5 hours to complete each test.

- Private Pilot—Airplane/Recreational Pilot–Transition
- Private Pilot—Helicopter/Recreational Pilot–Transition
- Private Pilot—Gyroplane/Recreational Pilot–Transition

The following tests each contain 60 questions, and you are allowed 2.5 hours to complete each test.

- Private Pilot—Airplane
- Private Pilot—Rotorcraft/Helicopter
- Private Pilot—Rotorcraft/Gyroplane
- Private Pilot—Glider
- Private Pilot—Balloon–Hot Air
- Private Pilot—Balloon–Gas
- Private Pilot—Lighter-Than-Air–Airship

Communication between individuals through the use of words is a complicated process. In addition to being an exercise in the application and use of aeronautical knowledge, a knowledge test is also an exercise in communication since it involves the use of the written language. Since the tests involve written rather than spoken words, communication between the test writer and the person being tested may become a difficult matter if care is not exercised by both parties. Consequently, considerable effort is expended to write each question in a clear, precise manner. Make sure you carefully read the instructions given with each test, as well as the statements in each test item.

When taking a test, keep the following points in mind:

- Answer each question in accordance with the latest regulations and guidance publications.
- Read each question carefully before looking at the possible answers. You should clearly understand the problem before attempting to solve it.
- After formulating an answer, determine which choice corresponds with that answer. The answer chosen should completely resolve the problem.
- From the answers given, it may appear that there is more than one possible answer; however, there is only one answer that is correct and complete. The other answers are either incomplete, erroneous, or represent common misconceptions.
- If a certain question is difficult for you, it is best to mark it for review and proceed to the next question. After you answer the less difficult questions, return to those which you marked for review and answer them. The review marking procedure will be explained to you prior to starting the test. Although the computer should alert you to unanswered questions, make sure every

question has an answer recorded. This procedure will enable you to use the available time to maximum advantage.

- When solving a calculation problem, select the answer closest to your solution. The problem has been checked with various types of calculators; therefore, if you have solved it correctly, your answer will be closer to the correct answer than any of the other choices.

Process for Taking a Knowledge Test

The Federal Aviation Administration (FAA) has available hundreds of computer testing centers worldwide. These testing centers offer the full range of airman knowledge tests including military competence, instrument foreign pilot, and pilot examiner predesignated tests. Refer to appendix 1 of this guide for a list of computer testing designees (CTD's).

The first step in taking a knowledge test is the registration process. You may either call the central 1-800 numbers (refer to appendix 1 for 1-800 numbers) or simply use the walk-in basis. If you choose to use the 1-800 number to register, you will need to select a testing center, schedule a test date, and make financial arrangements for test payment. You may register for tests several weeks in advance, and you may cancel your appointment according to the CTD's cancellation policy. If you do not follow the CTD's cancellation policies, you could be subject to a cancellation fee.

The next step in taking a knowledge test is providing proper identification. You should determine what knowledge test prerequisites are necessary before going to the computer testing center. Your instructor or local Flight Standards District Office (FSDO) can assist you with what documentation to take to the testing facility. Testing center personnel will not begin the test until your identification is verified. A limited number of tests do not require authorization.

Acceptable forms of authorization:

- A certificate of graduation or a statement of accomplishment certifying the satisfactory completion of the ground school portion of a course from an FAA-certificated pilot school.

- A certificate of graduation or a statement of accomplishment certifying the satisfactory completion of the ground school portion of a course from an agency such as a high school, college, adult education program,

- U.S. Armed Force, ROTC Flight Training School, or Civil Air Patrol.

- A written statement or logbook endorsement from an authorized instructor certifying that you have accomplished a ground training or home study course required for the rating sought and you are prepared for the knowledge test.

- Failed Airman Test Report, passing Airman Test Report, or expired Airman Test Report (pass or fail), provided that you still have the original Airman Test Report in your possession.

Before you take the actual test, you will have the option to take a sample test. The actual test is time limited; however, you should have sufficient time to complete and review your test.

Upon completion of the knowledge test, you will receive your Airman Test Report, with the testing center's embossed seal, which reflects your score.

The Airman Test Report lists the subject matter knowledge codes for questions answered incorrectly. The total number of subject matter knowledge codes shown on the Airman Test Report is not necessarily an indication of the total number of questions answered incorrectly. Appendix 1 contains a list of subject matter knowledge codes that refer to the knowledge areas. Study these knowledge areas to improve your understanding of the subject matter.

Your instructor is required to provide instruction on each of the knowledge areas listed on your Airman Test Report and to complete an endorsement of this instruction. The Airman Test Report must be presented to the examiner prior to taking the practical test. During the oral portion of the practical test, the examiner is required to evaluate the noted areas of deficiency.

Should you require a duplicate Airman Test Report due to loss or destruction of the original, send a signed request accompanied by a check or money order for $1 payable to the FAA. Your request should be sent to the Federal Aviation Administration, Airmen Certification Branch, AFS-760, P.O. Box 25082, Oklahoma City, OK 73125.

USE OF TEST AIDS AND MATERIALS

Airman knowledge tests require applicants to analyze the relationship between variables needed to solve aviation problems, in addition to testing for accuracy of a mathematical calculation. The intent is that all applicants are tested on concepts rather than rote calculation ability. It is permissible to use certain calculating devices when taking airman knowledge tests, provided they are used within the following guidelines. The term "calculating devices" is interchangeable with such items as calculators, computers, or any similar devices designed for aviation-related activities.

1. Guidelines for use of test aids and materials. The applicant may use test aids and materials within the guidelines listed below, if actual test questions or answers are not revealed.

a. Applicants may use test aids, such as scales, straightedges, protractors, plotters, navigation computers, log sheets, and all models of aviation-oriented calculating devices that are directly related to the test. In addition, applicants may use any test materials provided with the test.

b. Manufacturer's permanently inscribed instructions on the front and back of such aids listed in 1(a), e.g., formulas, conversions, regulations, signals, weather data, holding pattern diagrams, frequencies, weight and balance formulas, and air traffic control procedures are permissible.

c. The test proctor may provide calculating devices to applicants and deny them use of their personal calculating devices if the applicant's device does not have a screen that indicates all memory has been erased. The test proctor must be able to determine the calculating device's erasure capability. The use of calculating devices incorporating permanent or continuous type memory circuits without erasure capability are prohibited.

d. The use of magnetic cards, magnetic tapes, modules, computer chips, or any other device upon which prewritten programs or information related to the test can be stored and retrieved are prohibited. Printouts of data will be surrendered at the completion of the test if the calculating device used incorporates this design feature.

e. The use of any booklet or manual containing instructions related to the use of the applicant's calculating device is not permitted.

f. Dictionaries are not allowed in the testing area.

g. The test proctor makes the final determination relating to test materials and personal possessions that the applicant may take into the testing area.

2. Guidelines for dyslexic applicant's use of test aids and materials. A dyslexic applicant may request approval from the local Flight Standards District Office (FSDO) to take an airman knowledge test using one of the three options listed in preferential order:

a. Option One. Use current testing facilities and procedures whenever possible.

b. Option Two. Applicants may use Franklin Speaking Wordmaster® to facilitate the testing process. The Wordmaster® is a self-contained electronic thesaurus that audibly pronounces typed in words and presents them on a display screen. It has a built-in headphone jack for private listening. The headphone feature will be used during testing to avoid disturbing others.

c. Option Three. Applicants who do not choose to use the first or second option may request a test proctor to assist in reading specific words or terms from the test questions and supplement material. In the interest of preventing compromise of the testing process, the test proctor should be someone who is non-aviation oriented. The test proctor will provide reading assistance only, with no explanation of words or terms. The Airman Testing Standards Branch, AFS-630, will assist in the selection of a test site and test proctor.

Cheating or Other Unauthorized Conduct

Computer testing centers must follow strict security procedures to avoid test compromise. These procedures are established by the FAA and are covered in FAA Order 8080.6, Conduct of Airman Knowledge Tests. The FAA has directed testing centers to terminate a test at any time a test proctor suspects a cheating incident has occurred. An FAA investigation will then be conducted. If the investigation determines that cheating or unauthorized conduct has occurred, then any airman certificate or rating that you hold may be revoked, and you will be prohibited for 1 year from applying for or taking any test for a certificate or rating under 14 CFR part 61.

Validity of Airman Test Reports

Airman Test Reports are valid for the 24-calendar month period preceding the month you complete the practical test. If the Airman Test Report expires before completion of the practical test, you must retake the knowledge test.

Retesting Procedures

If you receive a grade lower than a 70 percent and wish to retest, you must present the following to testing center personnel.

- failed Airman Test Report; and
- a written endorsement from an authorized instructor certifying that additional instruction has been given, and the instructor finds you competent to pass the test.

If you decide to retake the test in anticipation of a better score, you may retake the test after 30 days from the date your last test was taken. The FAA will not allow you to retake a passed test before the 30-day period has lapsed. Prior to retesting, you must give your current Airman Test Report to the test proctor. The last test taken will reflect the official score.

SAMPLE TEST QUESTIONS AND ANSWERS

RECREATIONAL PILOT—AIRPLANE (RPA)

1. A recreational pilot acting as pilot in command must have in his/her personal possession while aboard the aircraft

A—a current logbook endorsement to show that a flight review has been satisfactorily accomplished.

B—the current and appropriate pilot and medical certificates.

C—the pilot logbook to show recent experience requirements to serve as pilot in command have been met.

Answer B—Subject Matter Knowledge Code: A29.

2. One of the main functions of flaps during approach and landing is to

A—decrease the angle of descent without increasing the airspeed.

B—permit a touchdown at a higher indicated airspeed.

C—increase the angle of descent without increasing the airspeed.

Answer C—Subject Matter Knowledge Code: H305.

3. A temperature inversion would most likely result in which weather condition?

A—Clouds with extensive vertical development above an inversion aloft.

B—Good visibility in the lower levels of the atmosphere and poor visibility above an inversion aloft.

C—An increase in temperature as altitude is increased.

Answer C—Subject Matter Knowledge Code: I21.

4. When telephoning a weather briefing facility for preflight weather information, pilots should

A—identify themselves as pilots.

B—tell the number of hours they have flown within the preceding 90 days.

C—state the number of occupants on board and the color of the aircraft.

Answer A—Subject Matter Knowledge Code: H320.

5. What action can a pilot take to aid in cooling an engine that is overheating during a climb?

A—Reduce rate of climb and increase airspeed.

B—Reduce climb speed and increase RPM.

C—Increase climb speed and increase RPM.

Answer A—Subject Matter Knowledge Code: H307.

RECREATIONAL PILOT—ROTORCRAFT/ HELICOPTER (RPH)

1. What exception, if any, permits a recreational pilot to act as pilot in command of an aircraft carrying a passenger for hire?

A—If the passenger pays no more than the operating expenses.

B—If a donation is made to a charitable organization for the flight.

C—There is no exception.

Answer C—Subject Matter Knowledge Code: A29.

2. The lift differential that exists between the advancing main rotor blade and the retreating main rotor blade is known as

A—transverse flow effect.

B—dissymmetry of lift.

C—hunting tendency.

Answer B—Subject Matter Knowledge Code: H71.

3. The amount of water vapor which air can hold depends on the

A—dewpoint.

B—air temperature.

C—stability of the air.

Answer B—Subject Matter Knowledge Code: I24.

4. When telephoning a weather briefing facility for preflight weather information, pilots should state the

A—full name and address of the pilot in command.

B—intended route, destination, and type of aircraft.

C—radio frequencies to be used.

Answer B—Subject Matter Knowledge Code: H320.

5. What action should the pilot take if engine failure occurs at altitude?

A—Open the throttle as the collective pitch is raised.

B—Reduce cyclic back stick pressure during turns.

C—Lower the collective pitch control, as necessary, to maintain rotor RPM.

Answer C—Subject Matter Knowledge Code: H80.

PILOT—ROTORCRAFT/GYROPLANE (RPG)

1. A recreational pilot may fly as sole occupant of an aircraft at night while under the supervision of a flight instructor provided the flight or surface visibility is at least

A—3 miles.

B—4 miles.

C—5 miles.

Answer C—Subject Matter Knowledge Code: A29.

2. What precaution should be taken while taxiing a gyroplane?

A—The cyclic stick should be held in the neutral position at all times.

B—Avoid abrupt control movements when blades are turning.

C—The cyclic stick should be held slightly aft of neutral at all times.

Answer B—Subject Matter Knowledge Code: H702.

3. What are characteristics of unstable air?

A—Turbulence and good surface visibility.

B—Turbulence and poor surface visibility.

C—Nimbostratus clouds and good surface visibility.

Answer A—Subject Matter Knowledge Code: I25.

4. When telephoning a weather briefing facility for preflight weather information, pilots should state

A—the full name and address of the formation commander.

B—that they possess a current pilot certificate.

C—whether they intend to fly VFR only.

Answer C—Subject Matter Knowledge Code: H320.

5. A below glide slope indication from a tri-color VASI is a

A—red light signal.

B—pink light signal.

C—green light signal.

Answer A—Subject Matter Knowledge Code: J03.

PRIVATE PILOT—AIRPLANE/ RECREATIONAL PILOT–TRANSITION (PAT)

1. In addition to other preflight actions for a VFR flight away from the vicinity of the departure airport, regulations specifically require the pilot in command to

A—review traffic control light signal procedures.

B—check the accuracy of the navigation equipment and the emergency locator transmitter (ELT).

C—determine runway lengths at airports of intended use and the aircraft's takeoff and landing distance data.

Answer C—Subject Matter Knowledge Code: B07.

2. While cruising at 9,500 feet MSL, the fuel/air mixture is properly adjusted. What will occur if a descent to 4,500 feet MSL is made without readjusting the mixture?

A—The fuel/air mixture may become excessively lean.B—There will be more fuel in the cylinders than is needed for normal combustion, and the excess fuel will absorb heat and cool the engine.

C—The excessively rich mixture will create higher cylinder head temperatures and may cause detonation.

Answer A—Subject Matter Knowledge Code: H307.

3. If a flight is made from an area of low pressure into an area of high pressure without the altimeter setting being adjusted, the altimeter will indicate

A—the actual altitude above sea level.

B—higher than the actual altitude above sea level.

C—lower than the actual altitude above sea level.

Answer C—Subject Matter Knowledge Code: I22.

4. (Refer to appendix 2, figure 1.) An aircraft departs an airport in the eastern daylight time zone at 0945 EDT for a 2-hour flight to an airport located in the centeral daylight time zone. The landing should be at what coordinated universal time?

A—1345Z.

B—1445Z.

C—1545Z.

Answer C—Subject Matter Knowledge Code: H340.

5. How is engine operation controlled on an engine equipped with a constant-speed propeller?

A—The throttle controls power output as registered on the manifold pressure gauge and the propeller control regulates engine RPM.

B—The throttle controls power output as registered on the manifold pressure gauge and the propeller control regulates a constant blade angle.

C—The throttle controls engine RPM as registered on the tachometer and the mixture control regulates the power output.

Answer A—Subject Matter Knowledge Code: H308.

PRIVATE PILOT—HELICOPTER/ RECREATIONAL PILOT–TRANSITION (PHT)

1. Under what conditions, if any, may a private pilot operate a helicopter under special VFR at night within Class D airspace?

A—The helicopter must be fully instrument equipped and the pilot must be instrument rated.

B—The flight visibility must be at least 1 mile.

C—There are no conditions; regulations permit this.

Answer C—Subject Matter Knowledge Code: B09.

2. (Refer to appendix 2, figure 2.) During flight, if cyclic control pressure is applied which results in a maximum increase in pitch angle of the rotor blade at position A, the rotor disc will tilt

A—forward.

B—aft.

C—left.

Answer A—Subject Matter Knowledge Code: H71.

3. One weather phenomenon which will always occur when flying across a front is a change in the

A—wind direction.

B—type of precipitation.

C—stability of the air mass.

Answer A—Subject Matter Knowledge Code: I27.

4. (Refer to appendix 2, figure 1.) An aircraft departs an airport in the central standard time zone at 0845 CST for a 2-hour flight to an airport located in the mountain standard time zone. The landing should be at what coordinated universal time?

A—1345Z.

B—1445Z.

C—1645Z.

Answer C—Subject Matter Knowledge Code: H340.

5. Which initial action should a pilot take prior to entering Class C airspace?

A—Contact approach control on the appropriate frequency.

B—Contact the tower and request permission to enter.

C—Contact the FSS for traffic advisories.

Answer A—Subject Matter Knowledge Code: J11.

PRIVATE PILOT—GYROPLANE/ RECREATIONAL PILOT–TRANSITION (PGT)

1. Except in Alaska, during what time period should lighted position lights be displayed on an aircraft?

A—End of evening civil twilight to the beginning of morning civil twilight.

B—1 hour after sunset to 1 hour before sunrise.

C—Sunset to sunrise.

Answer C—Subject Matter Knowledge Code: B11.

2. If the pilot experiences ground resonance during rotor spin-up, what action should the pilot take?

A—Taxi to a smooth area.

B—Close the throttle and slowly raise the spin-up lever.

C—Make a normal takeoff immediately.

Answer B—Subject Matter Knowledge Code: H709.

3. Individual forecasts for specific routes of flight can be obtained from which weather source?

A—Transcribed Weather Broadcasts (TWEB's).

B—Terminal Forecasts.

C—Area Forecasts.

Answer A—Subject Matter Knowledge Code: I40.

4. (Refer to appendix 2, figure 1.) An aircraft departs an airport in the Pacific standard time zone at 1030 PST for a 4-hour flight to an airport located in the central standard time zone. The landing should be at what coordinated universal time?

A—2030Z.

B—2130Z.

C—2230Z.

Answer C—Subject Matter Knowledge Code: H340.

5. When taking off or landing at an airport where heavy aircraft are operating, one should be particularly alert to the hazards of wingtip vortices because this turbulence tends to

A—rise from a crossing runway into the takeoff or landing path.

B—rise into the traffic pattern area surrounding the airport.

C—sink into the flightpath of aircraft operating below the aircraft generating the turbulence.

Answer C—Subject Matter Knowledge Code: J27.

PRIVATE PILOT—AIRPLANE (PAR)

1. The three takeoffs and landings that are required to act as pilot in command at night must be done during the time period from

A—sunset to sunrise.

B—1 hour after sunset to 1 hour before sunrise.

C—the end of evening civil twilight to the beginning of morning civil twilight.

Answer B—Subject Matter Knowledge Code: A20.

2. In what flight condition is torque effect the greatest in a single-engine airplane?

A—Low airspeed, high power, high angle of attack.

B—Low airspeed, low power, low angle of attack.

C—High airspeed, high power, high angle of attack.

Answer A—Subject Matter Knowledge Code: H300.

3. The wind at 5,000 feet AGL is southwesterly while the surface wind is southerly. This difference in direction is primarily due to

A—stronger pressure gradient at higher altitudes.

B—friction between the wind and the surface.

C—stronger Coriolis force at the surface.

Answer B—Subject Matter Knowledge Code: I23.

4. (Refer to appendix 2, figure 3, illustration 1.) The VOR receiver has the indications shown. What is the aircraft's position relative to the station?

A—North.

B—East.

C—South.

Answer C—Subject Matter Knowledge Code: H348.

5. The most effective method of scanning for other aircraft for collision avoidance during nighttime hours is to use

A—regularly spaced concentration on the 3-, 9-, and 12-o'clock positions.

B—a series of short, regularly spaced eye movements to search each 30-degree sector.

C—peripheral vision by scanning small sectors and utilizing offcenter viewing.

Answer C—Subject Matter Knowledge Code: H63.

PRIVATE PILOT—ROTORCRAFT/ HELICOPTER (PRH)

1. No person may begin a flight in a rotorcraft under VFR unless there is enough fuel to fly to the first point of intended landing and, assuming normal cruising speed, to fly thereafter for at least

A—20 minutes.

B—30 minutes.

C—1 hour.

Answer A—Subject Matter Knowledge Code: B09.

2. During forward cruising flight at constant airspeed and altitude, the individual rotor blades, when compared to each other, are operating

A—with increased lift on the retreating blade.

B—with a decreasing angle of attack on the advancing blade.

C—at unequal airspeed, unequal angles of attack, and equal lift moment.

Answer C—Subject Matter Knowledge Code: H71.

3. Which conditions result in the formation of frost?

A—The temperature of the collecting surface is at or below freezing when small droplets of moisture fall on the surface.

B—The temperature of the collecting surface is at or below the dewpoint of the adjacent air and the dewpoint is below freezing.

C—The temperature of the surrounding air is at or below freezing when small drops of moisture fall on the collecting surface.

Answer B—Subject Matter Knowledge Code: I24.

4. (Refer to appendix 2, figure 3, illustration 3.) The VOR receiver has the indications shown. What is the aircraft's position relative to the station?

A—East.

B—Southeast.

C—West.

Answer B—Subject Matter Knowledge Code: H348.

5. Under what condition, if any, may pilots fly through a restricted area?

A—When flying on airways with an ATC clearance.

B—With the controlling agency's authorization.

C—Regulations do not allow this.

Answer B—Subject Matter Knowledge Code: J09.

PRIVATE PILOT—ROTORCRAFT/ GYROPLANE (PRG)

1. A special VFR clearance authorizes the pilot of an aircraft to operate VFR while within Class D airspace when the visibility is

A—less than 1 mile and the ceiling is less than 1,000 feet.

B—at least 1 mile and the aircraft can remain clear of clouds.

C—at least 3 miles and the aircraft can remain clear of clouds.

Answer B—Subject Matter Knowledge Code: B09.

2. High airspeeds, particularly in turbulent air, should be avoided primarily because of the possibility of

A—an abrupt pitchup.

B—retreating blade stall.

C—a low-frequency vibration developing.

Answer B—Subject Matter Knowledge Code: H78.

3. What measurement can be used to determine the stability of the atmosphere?

A—Atmospheric pressure.

B—Actual lapse rate.

C—Surface temperature.

Answer B—Subject Matter Knowledge Code: I25.

4. (Refer to appendix 2, figure 3, illustration 8.) The VOR receiver has the indications shown. What radial is the aircraft crossing?

A—030.

B—210.

C—300.

Answer A—Subject Matter Knowledge Code: H348.

5. VFR approaches to land at night should be accomplished

A—at a higher airspeed.

B—with a steeper descent.

C—the same as during daytime.

Answer C—Subject Matter Knowledge Code: H63.

PRIVATE PILOT—GLIDER (PGL)

1. The minimum allowable strength of a towline used for an aerotow of a glider having a certificated gross weight of 700 pounds is

A—560 pounds.

B—700 pounds.

C—1,000 pounds.

Answer A—Subject Matter Knowledge Code: B12.

2. A sailplane has a best glide ratio of 23:1. How many feet will the glider lose in 8 nautical miles?

A—1,840 feet.

B—2,100 feet.

C—2,750 feet.

Answer B—Subject Matter Knowledge Code: N27.

3. What cloud types would indicate convective turbulence?

A—Cirrus clouds.

B—Nimbostratus clouds.

C—Towering cumulus clouds.

Answer C—Subject Matter Knowledge Code: I26.

4. When telephoning a weather briefing facility for preflight weather information, pilots should

A—identify themselves as pilots.

B—tell the number of hours they have flown within the preceding 90 days.

C—state the number of occupants on board and the color of the aircraft.

Answer A—Subject Matter Knowledge Code: H526.

5. What minimum upward current must a glider encounter to maintain altitude?

A—At least 2 feet per second.

B—The same as the glider's sink rate.

C—The same as the adjacent down currents.

Answer B—Subject Matter Knowledge Code: I35.

PRIVATE PILOT—BALLOON–HOT AIR (PBH)

1. Prior to becoming certified as a private pilot with a balloon rating, the pilot must have at least

A—passed a third-class medical exam.

B—obtained a statement from a designated medical examiner.

C—made a statement certifying that he/she has no known medical deficiency that would make him/her unable to act as pilot.

Answer C—Subject Matter Knowledge Code: A23.

2. If ample propane is available, within which temperature range will propane vaporize sufficiently to provide enough pressure for burner operation during flight?

A—0 to 30 °F.

B—10 to 30 °F.

C—30 to 90 °F.

Answer C—Subject Matter Knowledge Code: O220.

3. The conditions necessary for the formation of cumulonimbus clouds are a lifting action and

A—unstable air containing an excess of condensation nuclei.

B—unstable, moist air.

C—either stable or unstable air.

Answer B—Subject Matter Knowledge Code: I30.

4. When telephoning a weather briefing facility for preflight weather information, pilots should state the

A—full name and address of the pilot in command.

B—intended route, destination, and type of aircraft.

C—radio frequencies to be used.

Answer B—Subject Matter Knowledge Code: H320.

5. When telephoning a weather briefing facility for preflight weather information, pilots should state

A—the full name and address of the formation commander.

B—that they possess a current pilot certificate.

C—whether they intend to fly VFR only.

Answer C—Subject Matter Knowledge Code: H05.

PRIVATE PILOT—BALLOON–GAS (PBG)

1. The person directly responsible for the pre-launch briefing of passengers for a flight is the

A—safety officer.

B—pilot in command.

C—ground crewmember.

Answer B—Subject Matter Knowledge Code: B07.

2. What is the relationship of false lift with the wind?

A—False lift increases as the wind accelerates the balloon.

B—False lift does not exist if the surface winds are calm.

C—False lift decreases as the wind accelerates the balloon.

Answer C—Subject Matter Knowledge Code: O30.

3. What conditions are necessary for the formation of thunderstorms?

A—High humidity, lifting force, and unstable conditions.

B—High humidity, high temperature, and cumulus clouds.

C—Lifting force, moist air, and extensive cloud cover.

Answer A—Subject Matter Knowledge Code: I30.

4. When telephoning a weather briefing facility for preflight weather information, pilots should state

A—the full name and address of the formation commander.

B—that they possess a current pilot certificate.

C—whether they intend to fly VFR only.

Answer C—Subject Matter Knowledge Code: H320.

5. The minimum size a launch site should be is at least

A—twice the height of the balloon.

B—100 feet for every 1 knot of wind.

C—500 feet on the downwind side.

Answer B—Subject Matter Knowledge Code: O30.

PRIVATE PILOT—LIGHTER-THAN-AIR–AIRSHIP (PLA)

1. An aircraft's annual inspection was performed on July 12, this year. The next annual inspection will be due no later than

A—July 1, next year.

B—July 13, next year.

C—July 31, next year.

Answer C—Subject Matter Knowledge Code: B13.

2. Under which condition will an airship float in the air?

A—When buoyant force equals horizontal equilibrium existing between propeller thrust and airship drag.

B—When buoyant force is less than the difference between airship weight and the weight of the air volume being displaced.

C—When buoyant force equals the difference between airship weight and the weight of the air volume being displaced.

Answer C—Subject Matter Knowledge Code: P01.

3. Ceiling is defined as the height above the Earth's surface of the

A—lowest reported obscuration and the highest layer of clouds reported as overcast.

B—lowest layer of clouds or obscuring phenomena reported as broken, overcast, and not classified as thin or partial.

C—lowest layer of clouds reported as scattered, broken, or thin.

Answer B—Subject Matter Knowledge Code: I41.

4. When the course deviation indicator (CDI) needle is centered during an omnireceiver check using a VOR test signal (VOT), the omnibearing selector (OBS) and the TO/FROM indicator should read

A—180° FROM, only if the pilot is due north of the VOT.

B—0° TO or 180° FROM, regardless of the pilot's position from the VOT.

C—0° FROM or 180° TO, regardless of the pilot's position from the VOT.

Answer C—Subject Matter Knowledge Code: J01.

5. Which takeoff procedure is considered to be most hazardous for an airship?

A—Maintaining only 50 percent of the maximum permissible positive angle of inclination.

B—Failing to apply full engine power properly on all takeoffs, regardless of wind.

C—Maintaining a negative angle of inclination during takeoff after elevator response is adequate for controllability.

Answer C—Subject Matter Knowledge Code: P11.

APPENDIX 1

LIST OF REFERENCE MATERIALS AND SUBJECT MATTER KNOWLEDGE CODES

The publications listed in the following pages contain study material you need to be familiar with when preparing for recreational pilot and private pilot knowledge tests. All of these publications can be purchased through U.S. Government Bookstores, commercial aviation supply houses, or industry organizations. The latest revision of the listed references should be requested. Additional study material is also available through these sources that may be helpful in preparing for recreational pilot and private pilot knowledge tests.

The subject matter knowledge codes refer to the specific reference for the knowledge standard. When reviewing results of your knowledge test, you should compare the subject matter knowledge code(s) on your Airman Test Report to the ones found below. This will be helpful for both review and preparation for the practical test.

Title 14 of the Code of Federal Regulations (14 CFR) part 1—Definitions and Abbreviations

A01	General Definitions
A02	Abbreviations and Symbols

14 CFR part 43—Maintenance, Preventive Maintenance, Rebuilding, and Alteration

A15	General
A16	Appendixes

14 CFR part 61—Certification: Pilots, Flight Instructors, and Ground Instructors

A20	General
A21	Aircraft Ratings and Pilot Authorizations
A22	Student Pilots
A23	Private Pilots

14 CFR part 71—Designation of Class A, Class B, Class C, Class D, and Class E Airspace Areas; Airways; Routes; and Reporting Points

A60	General—Class A Airspace

14 CFR part 91—General Operating and Flight Rules

B07	General
B08	Flight Rules—General

B09	Visual Flight Rules
B11	Equipment, Instrument, and Certification Requirements
B12	Special Flight Operations
B13	Maintenance, Preventive Maintenance, and Alterations

NTSB 830—Rules Pertaining to the Notification and Reporting of Aircraft Accidents or Incidents and Overdue Aircraft, and Preservation of Aircraft Wreckage, Mail, Cargo, and Records

G10	General
G11	Initial Notification of Aircraft Accidents, Incidents, and Overdue Aircraft
G12	Preservation of Aircraft Wreckage, Mail, Cargo, and Records
G13	Reporting of Aircraft Accidents, Incidents, and Overdue Aircraft

AC 61-23—Pilot's Handbook of Aeronautical Knowledge

H300	Forces Acting on the Airplane in Flight
H301	Turning Tendency (Torque Effect)
H302	Airplane Stability
H303	Loads and Load Factors
H304	Airplane Structure
H305	Flight Control Systems
H306	Electrical System
H307	Engine Operation
H308	Propeller
H309	Starting the Engine
H310	Exhaust Gas Temperature Gauge
H311	Aircraft Documents, Maintenance, and Inspections
H312	The Pitot-Static System and Associated Instruments
H313	Gyroscopic Flight Instruments
H314	Magnetic Compass
H315	Weight Control
H316	Balance, Stability, and Center of Gravity
H317	Airplane Performance
H318	Observations
H319	Service Outlets
H320	Weather Briefings

H321	Nature of the Atmosphere
H322	The Cause of Atmospheric Circulation
H323	Moisture and Temperature
H324	Air Masses and Fronts
H325	Aviation Weather Reports, Forecasts, and Weather Charts
H326	Types of Airports
H327	Sources for Airport Data
H328	Airport Markings and Signs
H329	Airport Lighting
H330	Wind Direction Indicators
H331	Radio Communications
H332	Air Traffic Services
H333	Wake Turbulence
H334	Collision Avoidance
H335	Controlled Airspace
H336	Uncontrolled Airspace
H337	Special Use Airspace
H338	Other Airspace Areas
H339	Aeronautical Charts
H340	Latitude and Longitude
H341	Effect of Wind
H342	Basic Calculations
H343	Pilotage
H344	Dead Reckoning
H345	Flight Planning
H346	Charting the Course
H347	Filing a VFR Flight Plan
H348	Radio Navigation
H349	Obtaining a Medical Certificate
H350	Health Factors Affecting Pilot Performance
H351	Environmental Factors which Affect Pilot Performance

AC 61-21—Flight Training Handbook

H50	Introduction to Flight Training
H52	Introduction to the Basics of Flight
H54	Ground Operations
H58	Landing Approaches and Landings
H60	Proficiency Flight Maneuvers
H63	Night Flying
H66	Principles of Flight and Performance Characteristics

AC 61-13—Basic Helicopter Handbook

H70	General Aerodynamics
H71	Aerodynamics of Flight
H73	Function of the Controls
H74	Other Helicopter Components and Their Functions
H76	Weight and Balance
H78	Some Hazards of Helicopter Flight
H79	Precautionary Measures and Critical Conditions
H80	Helicopter Flight Maneuvers
H81	Confined Area, Pinnacle, and Ridgeline Operations

Gyroplane Training Manual—Graves Publishing Co.

H95	General

Understanding the Gyroplane—The Abbott Co.

H650	Magic of Rotor Blades
H651	Behind the Power Curve
H652	Beating P.I.O.

Gyroplane Flight Training Manual—Jean-Pierre Harrison

H701	General Aerodynamics
H702	Aerodynamics of Flight
H704	Function of the Controls
H709	Some Hazards of Gyroplane Flight
H710	Precautionary Measures and Critical Conditions
H711	Gyroplane Flight Maneuvers

AC 00-6—Aviation Weather

I21	Temperature
I22	Atmospheric Pressure and Altimetry
I23	Wind
I24	Moisture, Cloud Formation, and Precipitation
I25	Stable and Unstable Air
I26	Clouds
I27	Air Masses and Fronts
I28	Turbulence
I29	Icing
I30	Thunderstorms
I31	Common IFR Producers
I33	Arctic Weather
I35	Soaring Weather
I36	Glossary of Weather Terms

AC 00-45—Aviation Weather Services

I40	The Aviation Weather Service Program
I41	Surface Aviation Weather Reports
I42	Pilot and Radar Reports and Satellite Pictures
I43	Aviation Weather Forecasts
I44	Surface Analysis Chart
I45	Weather Depiction Chart
I46	Radar Summary Chart
I47	Significant Weather Prognostics

AIM —Aeronautical Information Manual

J01	Air Navigation Radio Aids
J03	Airport Lighting Aids
J05	Airport Marking Aids and Signs
J08	Controlled Airspace
J09	Special Use Airspace
J10	Other Airspace Areas
J11	Service Available to Pilots
J12	Radio Communications Phraseology and Techniques

J13	Airport Operations
J14	ATC Clearance/Separations
J15	Preflight
J22	Emergency Services Available to Pilots
J25	Meteorology
J27	Wake Turbulence
J31	Fitness for Flight
J33	Pilot Controller Glossary

Other Documents

J34	Airport/Facility Directory
J37	Sectional Chart

AC 67-2—Medical Handbook for Pilots

J52	Hypoxia
J53	Hyperventilation
J58	Carbon Monoxide
J60	Night Flight
J61	Cockpit Lighting
J62	Disorientation (Vertigo)

Additional Advisory Circulars

M52	AC 00-2, Advisory Circular Checklist

Soaring Flight Manual—Jeppesen-Sanderson, Inc.

N20	Sailplane Aerodynamics
N21	Performance Considerations
N22	Flight Instruments
N23	Weather for Soaring
N24	Medical Factors
N25	Flight Publications and Airspace
N26	Aeronautical Charts and Navigation
N27	Computations for Soaring
N28	Personal Equipment
N29	Preflight and Ground Operations
N30	Aerotow Launch Procedures

N31	Ground Launch Procedures
N32	Basic Flight Maneuvers and Traffic
N33	Soaring Techniques
N34	Cross-Country Soaring

Flight Instructor Manual—Balloon Federation of America

O10	Flight Instruction Aids
O11	Human Behavior and Pilot Proficiency
O12	The Flight Check and the Designated Examiner

Balloon Digest—Balloon Federation of America

O150	Balloon—Theory and Practice
O155	Structure of the Modern Balloon
O160	Lift-off to Landing
O165	Weather for the Balloonist
O170	Propane and Fuel Management
O171	Chemical and Physical Properties
O172	Tanks
O173	Burners
O174	Hoses
O175	Refueling
O176	Fuel Contamination
O177	Heat Tapes (Coils)
O178	Nitrogen Pressurization
O179	Repairs and Maintenance

Powerline Excerpts—Balloon Federation of America

O30	Excerpts

Balloon Ground School—Balloon Publishing Co.

O220	Balloon Operations

How To Fly A Balloon—Balloon Publishing Co.

O250	Basic Terminology
O251	History
O252	Physics
O253	Equipment

O254	Checklists
O255	Flight Planning
O256	Preflight Operations
O257	The Standard Burn
O258	Inflation
O259	Launch
O260	Level Flight
O261	Ascents and Descents
O262	Contour Flying
O263	Maneuvering
O264	Approach to Landing
O265	Landings
O266	Deflation
O267	The Chase
O268	Landowners Relations
O269	Recovery and Pack-up
O270	Propane: Management and Fueling
O271	Tethering
O272	Emergency Procedures
O273	Skill Development
O274	Crew
O275	What is a Good Instructor
O276	Regulations
O277	Maintenance
O278	Earning a Pilot Certificate
O279	Radio Communications
O280	Appendix 1: Glossary

Goodyear Airship Operations Manual

P01	Buoyancy
P03	Free Ballooning
P04	Aerostatics

FAA Accident Prevention Program Bulletins

V01	FAA-P-8740-2, Density Altitude
V02	FAA-P-8740-5, Weight and Balance
V03	FAA-P-8740-12, Thunderstorms
V04	FAA-P-8740-19, Flying Light Twins Safely
V05	FAA-P-8740-23, Planning Your Takeoff
V06	FAA-P-8740-24, Tips on Winter Flying
V07	FAA-P-8740-25, Always Leave Yourself an Out
V08	FAA-P-8740-30, How to Obtain a Good Weather Briefing
V09	FAA-P-8740-40, Wind Shear
V10	FAA-P-8740-41, Medical Facts For Pilots
V11	FAA-P-8740-44, Impossible Turns
V12	FAA-P-8740-48, On Landings, Part I
V13	FAA-P-8740-49, On Landings, Part II
V14	FAA-P-8740-50, On Landings, Part III
V15	FAA-P-8740-51, How to Avoid a Midair Collision
V16	FAA-P-8740-52, The Silent Emergency

NOTE: AC 00-2, Advisory Circular Checklist, transmits the status of all FAA advisory circulars (AC's), as well as FAA internal publications and miscellaneous flight information, such as Aeronautical Information Manual, Airport/Facility Directory, practical test standards, knowledge test guides, and other material directly related to a certificate or rating. AC 00-2 is accessible through the Internet at http://www.faa.gov/abc/ac-chklst/actoc.htm, or you may obtain a free copy from:

U.S. Department of Transportation
Subsequent Distribution Office, SVC-121.23
Ardmore East Business Center
3341 Q 75 Ave.
Landover, MD 20785

COMPUTER TESTING DESIGNEES

The following is a list of the computer testing designees authorized to give FAA airman knowledge tests. This list should be helpful in case you choose to register for a test or simply want more information.

Computer Assisted Testing Service (CATS)
1849 Old Bayshore Highway
Burlingame, CA 94010

Applicant inquiry and test registration: 1-800-947-4228
From outside the U.S. (650) 259-8550

Sylvan Prometric
1000 Lancaster Street
Baltimore, MD 21202
Applicant inquiry and test registration: 1-800-274-1900, 1-800-967-1100, or 1-800-359-3278
From outside the U.S. registrants should contact the appropriate Regional Service Center (RSC):

London, England RSC	44-181-607-9090
Paris, France RSC	33-1-4289-3122
Dusseldorf, Germany RSC	49-2159-9233-50
Tokyo, Japan RSC	813-3269-9620
Latin America RSC	(612) 820-5200

LaserGrade Computer Testing
16209 S.E. McGillivray, Suite L
Vancouver, WA 98683
Applicant inquiry and test registration: 1-800-211-2753 or 1-800-211-2754
From outside the U.S. (360) 896-9111